David Bowen's do-it-yourself
weather guide...

TELL YOUR OWN WEATHER

wherever you are in Britain use
this weather guide to tell you
what is likely to follow

© David Bowen 2016

1

PART ONE
the observient

EYE AND A NOSE FOR THE WEATHER

those who are weatherwise are seldom otherwise

Telling the weather ...

There are two parts to this weather guide, the second far from guaranteed but hopefully with a laugh or two to make up and grains of wisdom now and then.

We have learned a lot from satellites since the time of my 'one-and six' *(7 decimal pence)* first edition of this weather guide of thirty pages in 1953, a period in which the Met Office forecasts have improved beyond measure, focus on individual regions and towns and are ready to give out special warnings. With a bit of luck we can plan for a week ahead, the job done for us, no more worries!

Well, not quite! To stay in touch one needs to be constantly making one's own check. There are times when a quick decision is needed to avoid the worst conditions. Best to play for safety when out and about, walking the hills or going to sea: – who knows where one will be in a week or even an hour from now? Sitting somewhere drenched? Or worse? A distant white cloud may prove to be a wolf in sheep's clothing, or if the wind gets up at a critical moment or a fog descends, what then? Did they predict it? Anyone know where we are?

As Tommy Cooper used to say, life is magic, *'just like that!'* It suddenly happens! And I haven't yet finished with him*!* - *b*ut, yes, 'just like that'... the weather particularly when what occurs is different from the overall picture. Why – one could be submerged in a clammy coastal fog while a mile or two inland it's bright sun and out comes the body cream. This kind of situation is typical of the British spring, with a cold sea.

As for longer-range forecasting, get it right once or twice and await laurels. A useful clue is that the longer a particular spell has lasted, the longer it may dominate – up to a point and probably not everywhere. Sages may know and you can be the sage!

David Bowen

CONTENTS   PART ONE

*(for Contents Part Two *see page 71)*

CONTENTS: PART ONE

1) Your tip from the clouds

Cumulus

Low-based clouds *(base average around 2,500 ft)*

> *If woolly fleeces spread the heavenly way*
> *Be sure no rain disturbs the sunny day...*

so long as the clouds are small, as shown above, with flat bases and curly, well- rounded, not ragged, tops that are slow-moving and neither growing vertically nor fast-spreading

Small Cumulus

and Stratocumulus

These are the individual small *cumulus* or else occurring in rolls or joined-up rows *(stratocumulus),* both above. Either type to be seen typically on settled days of spring and summer where a limited amount of cloud develops

They are formed by air currents rising over the heated ground if only to a limited extent in a 'stable' atmosphere. If they rush across the sky and enlarge or lead to overcast, it is a different situation and the weather is on the change

A lower, sometimes uniform grey base, or streaky layers, could be associated with fog and will readily brush hill-tops or submerge them. They indicate fine weather only if seen rising or otherwise dispersing soon after dawn; bad weather if they sink, persist or amalgamate

When showers are on the menu...

Very often the atmosphere is not as stable as one might assume from the day's beginning. One may get the impression of fine weather at the start only to find it different altogether after an hour or more as clouds grow ever larger. Gone the greeting of a beautiful white creation with the sun on it. Increasing upward growth and horizontal spread will have cast shadows and put paid to any early hope of warmth.

This will be a day of rising west to north-west wind, one that, inland, anyway, began relatively smoothly, almost calm. With luck there may be some 'creative' sunny periods to come; without it, mostly a 'draughty' wind and fiercely dominant clouds already towering into the sky – cumulonimbus, the largest in the cumulus family

Large Cumulus

Cumulonimbus

When the sky is full of rocks and owers
The earth's refreshed by frequent showers...

or as Shakespeare put it in 'Antony and Cleopatra',

Sometimes we see a cloud that's dragonish...
A vapour like a bear or lion,
Towered citadel or pendent rock ...

The cloud is in turmoil, its currents rising to a great height and resulting in spectacular displays. Within them there are wilder winds than any experienced over the ground. Sides and tops of clouds may merge while the landscape gets darker. Thunder and lightning are possible. Showers will saturate higher ground to windward and across exposed coasts backed by hills that force the wind to rise

They may begin quite early and become increasingly frequent. The best of the weather will occur well inland, especially to leeward of mountains

Note. As so often, the weather doesn't 'do things by halves'. Cumulonimbus clouds that are monsters in themselves may amalgamate so that the tops become 'out of shape', no more anvils to be seen but an over topping area of white if the sun is on it and observed impressively from a distance

While 'cu-nimb' prevails on cool showery days, it is also present at either the interruption or the break of a hot spell but, then, mostly submerged by haze.   Be not deceived! Here, though, cloud movement will be slower unless a wind arises to indicate that a change of weather-type is taking place

Final word on 'cu-nimb' from Lucretius 99-55 B.C.

Now clouds combine and spread o'er all the sky
When little ragged parts ascend on high -
Which may be twined, though by a feeble tie-
To make small clouds which, driven by the wind
To other like and little clouds are joined;
And these increase by more until they form
Thick heavy clouds, from whence proceeds a storm
Creech's 'Lucretius'

Altocumulus

and Altostratus

**Middle-height cloud (base average 8,000-10,000 ft -
prefix alto- either altocumulus or altostratus, namely...**

1)   **altocumulus**, a pattern of white, 'saucers', like small
cumulus only higher and flatter.   These usually indicate no
immediate or radical change for a while

2)   **altocumulus castellanus**, like 'judges' wigs' or lofty
miniature 'castles' where there are compact growing sides which
point to thundery conditions.   These formations are again
relatively small, like viewing an elevated chessboard, so that they
are easily missed and may fail to 'hit the eye'

3)   **altostratus**, a barely transparent greyish-white sheet, base
first around l5,000 ft, gradually lowering, almost invariably a sign
of rain and wind, cool in summer although mild in winter as
Atlantic weather reaches north-west -Europe...but, by contrast,
resulting in snowfall when hard-pressed against a bank of cold
Continental air already in position.   Ensuing rain from this
formation will last for several hours until typically giving way to
broken skies and showers

High cloud (cirrus, cirrocumulus, cirrostratus, average base 18, 000 – 25,000 ft)

As well as those that are lower, these distinctive formations embrace different types of weather. They mostly give long notice of their intentions. In fine weather small broken threads of cirrus indicate no immediate change, but ...

> *Trace in the sky the painter's brush*
> *And winds around you soon will rush*

Or (to put it another way,

> *Hens' scarts and mares' tails*
> *Make tall ships carry low sails*

a storm is on the way!

Mare's Tails

Cirrus fragments

Cirrus ('mares' 'tails') to cirrostratus

Within the hour the *cirrus* tails or hooks will have been replaced by an all-over white gauze across the entire horizon, This thin, milky high cloud sheet *cirrostratus* is less dense than altostratus but soon to be replaced by this. The outline of the sun or moon will shortly be obscured after first appearing as a halo, then, later, as though viewed through ground glass...

The forecast is clear enough. The weather is fast deteriorating. In technical terms a 'warm front' of a 'low' (low pressure system) is approaching, winds will freshen from, probably, the south-west with rain likely within an hour or so

The cloudsheet, all the time thickening, will eventually achieve a low base, about 2,000 ft, nimbostratus - remaining thus for several hours until an eventual part-clearance with some lighter rain or drizzle or to broken skies with showers

Cirrus increasing (seen against the sun)

Cirrocumulus .... 'mackerel' sky

Cirrocumulus, or by its more familiar name, mackerel sky, above, is an eye-catching and enchanting formation similar to altocumulus (as previously described) but with smaller globules and little or no shadowing. It may occur on the edge of a low pressure system that is at a distance and moving at an angle to the observer rather than directly towards

There should not be any immediate change in the weather , but the almanac-makers of old liked to err on the side of caution, thus ...

17

Mackerel sky, mackerel sky,
Never long wet and never long dry

Useful sayings about he clouds...

a) According to colour and outline

Light, delicate, quiet tints or colours, or soft, undefined forms of cloud indicate and accompany fine weather ; but unusual, gaudy hues in clouds of a hard outline foretell rain and, probably, strong wind with little notice being given

b) Capping hills

When mountains and hills appear capped by clouds that hang about and embrace them, storms are imminent

c) Bank in west

A bench (or bank) of cloud in the west foretells rain

d) Stationary, piling up

When clouds are stationary and others accumulate by them (but the first remain still) it is the sign of a storm

e) Settling back

When a heavy cloud comes up in the south-west and seems to settle back again, look out for a storm

f) Soon collecting

If the morning sky, from being clear, becomes quickly fretted with bunches of clouds, rain will soon fall

g) A curdly sky

Will not leave the earth long dry

h) Moving in different directions

If two layers of cloud appear in hot weather to begin moving in different directions, they indicate thunder

i) From the south, during a frost

If, during a frost, clouds drive up high from the south, expect a thaw

j) Early disappearance

When at sunrise the clouds are driven away, fair weather will follow

k) Clouds overhead, or otherwise

When it is bright all round it will not rain; when it is bright only overhead, it will

Bright all round

Cloud tints

The evening red and the morning grey
Are the tokens of a bonny day

(the 'morning grey' in this case being the close of a foggy night inland....but 'red sky in the morning,- shepherd's warning')...

A red morn... ever yet betokened
Wreck to the seaman, tempest to the field,
Sorrow to shepherds, woe unto the birds,
Gusts and foul flaws to herdmen and to herds

a) Zinc-grey layer

A low zinc-grey layer covering a wintry sky is a sign of snow. For large flakes the temperature may be above freezing

b) Green

If the clouds take on a dull greenish hue, heavy rain is imminent.

c) Milky

If the sky looks washed with a milky white
Rain is approaching if not yet in sight

d) Blue sky, no clouds

A clear blue sky without cloud is only to be trusted if the barometer is high and the reading steady

e) Mixed colours

A cloudy evening marked by a variety of colours, through orange and yellow to red and purple tints, is ever a sign of unsettled weather past, present and, probably, future

****

Statistics ... *interested?*

Just how much water lies over our heads within a single cloud?

Sir Graham Sutton, former Director of the Met. Office calculated that this could amount to 100,000 tons for a single shower cloud. That of a storm area would astound the imagination

Rain falling thickly from a cloudbase

2) Feel a wind take the rise?

Battle stations

a) When winds are light

Inland, the smaller and lighter winds generally rise in the morning and fall towards night

b) Sudden gusts

Sudden gusts are rare in a clear sky but mainly when it is cloudy and with rain

c) Suddenly freshening

The sudden storm lasts not three hours

d) Veering and backing

> *A veering wind, fair weather,*
> *A backing wind, foul weather*

(a veer being a clockwise change of direction, so typically in Britain from south to south-west or south-west to west

e) Wind with rain (nautical)

> *When the rain comes before the wind*
> *Halyards, sheets and braces mind;*
> *When he wind comes before the rain*
> *Soon you may set sail again*

f) With frost

When the hoar-frost* is accompanied by east wind, the cold will continue for a long time

*water vapour deposited and frozen in clear weather

g) North-west

> *A north-westerly gale*
> *Brings showers of hail*

h) North-west compared with north-east

Short storms succeeded by sunny intervals are the hall-mark of a brisk north-west wind, longer storms and more persistent cloud with one from north-east

i) Northerly

Northerly winds bring mostly showers rather than continuous rain or snow, *but...*

> *The north wind doth blow*
> *And we shall have snow*...*

*Snow flurries or showers. For continuous snow the wind is frequently between south-east and east during winter, and occasionally between south and south-east where there is continental air to clash with that from the Atlantic

j) West

A western wind carrieth water in his hand

Most of Britain's rain comes with a wind from between south-west and north-west

k) South, in summer

The south wind, when gentle, is not a great collector of clouds during the summer months, but if it becomes violent it quickly turns the sky cloudy and brings thunder and rain

l) South, in winter

The south wind during the winter months will bring mild, cloudy weather with, typically, drizzle or patchy rainfall

m) East

With a wind from between east and south-east come Britain's highest temperatures in summer but also the lowest in winter

With a high barometer it will stay dry – but if this starts to fall thundery rains occur in summer, and, in winter, snow that will only turn to rain when the wind veers to south or south-west

n) Unsteady

Unsteadiness of wind shows changing weather

o) Frequent changes

Frequent changes of direction and of variation in speed point to an established pattern of generally unsettled weather ; the more so when clouds are agitated

p) Making a swell

> *When there's hardly a wind but a*
> *swell on the sea*
> *For certain a drench and a gale*
> *there will be*

A capful of wind

3) Now you see it, now you don't!

a) On a clear day

... one may view distant hills at near on 50 miles distance and what is visible between. Wind likely to be moderate west to north-west, perhaps with some showers. Too much rain will reduce the sight

b) On a settled weather day

... visibility will be no more than moderate. As wind strength decreases haze will collect in the lower atmosphere - typically, with a south-east drift from the Continent to darken the sky.

c) The ideal?

Days that in summer prove ideal for the garden or beach will be poor for the landscape photographer. Visibility will reduce drastically...

...and here in Britain we may manufacture our own haze, mist or fog as well as import it from the Continent. In the past, before the control of smoke emission, large towns suffered from 'pea-soup'- fogs, the last notable ones those of London in the Decembers of 1953 and 1963

Swirling fog

Mist (caused by minute water droplets)

a) From the sea

When the mist comes from the sea
Then good weather it will be

(typically, between spring and early summer along coasts while it it may be sunny just a few miles inland)

b) Rising, falling or persisting over hills

White, fleecy, morning mist, slowly ascending high ground indicates a fine day to follow - the reverse if descending, persisting or increasing at this time

Mist clinging to sides of hill(s)

c) Evening mist

Patchy mist forming late after a fine day in the hills may be expected to rise and disappear next day

d) Rising from low ground

Overnight mist rising from low ground will soon vanish, to give way to a sunny day – the exception being in winter or late year when it may persist

e) With wind

Expect rain to follow if mist and wind arise together or whenever the two combine

f) Mists at sea and over land, according to season

Sea mists tend to be persistent whenever the ocean is cold and the wind absent or flowing from south or east; land mist always a risk when the wind drops and temperatures are low

Early morning mist

Haze (mostly man-made, reducing visibility)

a) In summer

The greater the haze, the more settled the weather

At the start of a fine spell between late spring and summer the weather will be clear to begin with but become hazy after a day or two and developing a thick haze before an eventual break in the weather when thunder-heads are seen to poke through

b) A thunderstorm will clear the air

True in respect to either a local and temporary heat storm or after a major thundery outbreak that signifies the end of a settled spell of weather. Not only will the atmosphere become much clearer, but it will also be cooler, at least for a while

   The query will then be centred on a possible return of fine weather or a more lasting spell of the cooler type that came in its wake. (This is where a barometer is useful – please see later section on barometric readings and changes, page 92)

Haze in winter

   In the colder spells of the winter, the wind (mainly from the east) it will be a case of Britain coming under a blanket of haze

   A veer in the wind from (usually) south-east to south or south-west will herald clearer weather after the passage of any initial snowfall. With a west to north wind the atmosphere stays clear

Hazy beach

4) Rain from Spain?...

... we just need to put up with it – as they did in Shakespeare's Twelfth Night 'with a hey, ho, the wind and the rain' (IV, i), so take it as it comes from wherever it comes ...or with experience and - dare one say, a spot of luck - determine how long it will last and what will follow ...

a) Rain early morning

Rain before seven
Clear by eleven

One may tend to rely on this, as it is more likely to be right than wrong, especially in the spring and early summer rather than later. It is really a 'rule of thumb' calculation, a tentative deduction based on the fact that our mostly westerly rainbelts generally clear within around five hours (some more rapidly).

b) Rain from the east

> *When the rain is from the east*
> *It lasts a day or two at least*

Because easterly winds tend to be persistent, any mixing of air at their fringe (which is often Britain and the immediate area) may be expected to continue for, perhaps, several days before the scene shifts and another weather regime takes its place. The easterly *unmixed* is associated with extreme temperatures in both summer and winter

c) Rain from the west and north-west

> *The western wind carrieth water in his hand*
> *Old proverb*

Most of the rainfall in Britain comes from the Atlantic, with a wind from between south-west and north-west if, strangely. only seldom from *due* west. The north-west gives powerfully gusty, showers, also thunder and hail ... a sudden coating of white, then gone. Hail in the wake of approximately half an hour of rain, wind and lightning marks the passage of a *cold front* and will lead to a temporary clearance with a temperature drop

d) Sudden, or with long warning

Sudden rains seldom last long, but when the air grows thick by degrees and the sun, moon and stars shine dimmer it shows that rain will set in and last for several hours

e) Light showers during a drought

If very light, short showers come during dry weather, they are said to *harden the drought* and indicate no general change

f) Heavy, long lasting

In summer, low pressure over southern Britain that is virtually stationary will be likely to produce more than even a single day's rain

g) In winter, followed by sudden frost

Sudden frosts in winter, after rain, soon bring back more rain again

h) Special events

Raining cats and dogs...

The often spirally fierce winds of a thunder-cloud may pick up frogs and other small creatures if not larger ones...

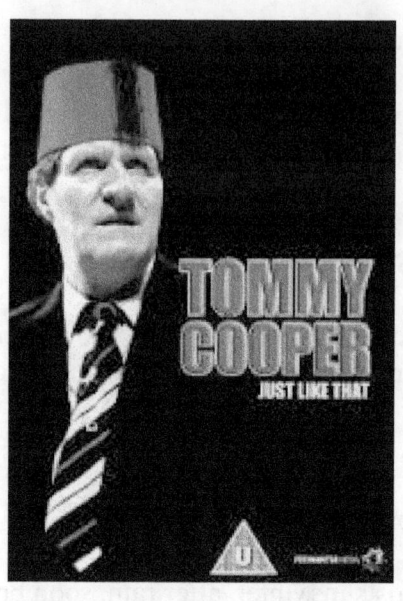

'just like that!

Flash floods – more of the weather"s theatricals!!

Sudden heavy rains of a local nature, can seldom be precisely forecast as to venue and may lead to 'flash flooding' in valleys. See the neatly parked car of one moment become the debris of the next and difficult to reach

5) ...Or where the rainbow ends?

a) Rainbow logic

Morning and night
> *A rainbow in the morning*
> *Is the shepherd's warning,*
> *A rainbow at night*
> *The shepherd's delight...*

so as for a red sky, so, one might conclude.   But not quite

The siting of rainbows, *above,* is within the context of a showery type of weather. The morning rainbow is likely to see just an immediate break of the sky between quite frequent showers; the evening rainbow an overnight break and clearing of the sky

b) Windward and leeward

A rainbow to windward, foul falls the day,
A rainbow to leeward, damp runs away

This saying can be used in conjunction with the previous one. It would be unusual for one to contradict the other

c) Suddenly disappearing

Where the clouds are not thick, a rainbow may form and vanish suddenly, or appear so to do with the prismatic colours being barely discernible. There may be a gradual improvement in the general weather at the time

6) Thundering weather!

'Look on my works, ye mighty, and despair'
P.B. Shelley, 'Ozymandias'

Significance of thunder

a) Much thunder

After much thunder, much rain

    The more prolonged the thunder, the greater the upset of the atmosphere and, therefore, not only plentiful rain to mark the occasion but also a pointer to an overall change in the weather from (normally) heat and settled conditions to an altogether more changeable and cooler style

41

b) Thunder from south
 or north

Thunder from the south or south-east denotes a long storm; from the north or north-west, a short storm or successive short storms well separated

c) Morning and noon

When it thunders in the morning it will rain before night. Thunder in the morning denotes winds; at noon, showers

d) In summer

Thunder and lightning in the summer show
The point from which the freshening wind will blow

A thunderstorm clears the air
It does so most effectively with a 'long innings' that precedes a general change in the weather

e) From different directions

Lightning may signify the approach of wind and rain from the quarter where it lightens but, if it lightens from different parts of the sky, prepare first for a long storm

f) With large or small falls of the barometer

A thunderstorm occurring with little or no fall of the barometer is unlikely to persist, but if the barometer falls there will be a longer storm and a general weather change

Lightning-prone areas and trees

Golf courses are among the areas most frequently struck by lightning, likewise beaches and lake-sides and high-positioned or isolated large trees

7) Feeling frosty today?...

Hard winters a rarity,
Snowfall a scarcity?
Spring frost a possibility!
.......be prepared!

Predict snow when frosty east winds are challenged by a south-east flow but resulting in a thaw if the process continues

Predict frost when...

1. Winds in winter settle from between east and north. Both day and night frost to be expected

2. Long-term while such winds persist

3. At night when winds in winter and spring from north-west to north and begin to fall

4   Any time in winter and in early to mid spring if the
    barometer reads 'Dry' or 'Very Dry' as a result of a 'High'
    positioned to the east of Britain that blocks the
    passage of milder air from the south or west

Predict a thaw, perhaps preceded by snow, when

1   In winter north to east winds give way to westerly
    or southerly, or with high cloud from the south

2   When the barometer falls gradually towards 'Change'
    or the sun takes on a watery look

8)   ... Or a mix of h. and c.?

When either human or the weather blows hot or cold there
has to be a reason...

Temperatures on trial

a) Falling

In spring sharp falls will occur towards and during the
night if the wind has been from north-west or north, when frost
must be expected, particularly downslope, in natural hollows

In summer any marked fall will be associated with a shift
of wind, probably from south-east to south-west, or from south-
west to north-west, with corresponding weather, rainy, or
showery, respectively

b) Rising quickly

A quick rise in temperature stands to be equally quickly
overturned, especially during summer

c) Rising slowly

Any rise in winter should be linked with the prevailing wind mild for south to south-west (not south-east). In summer a slow rise could indicate the start of fine weather of normally up to a week's duration, although sometimes longer

d) Damp summer heat

Towards the end of a hot sunny spell the atmosphere may feel oppressive as humidity rises. When this continues day after day, expect thundery outbreaks of rain, at least, locally, but eventually signifying a general change to cooler conditions.

9) Hear things?...

Sounds that carry

a) Good hearing day

This is likely to be caused when the atmosphere is moist in its lower levels with little or no wind. Sound will carry far in such conditions, as indeed it will in fog and mist. The bark of a distant dog can then travel for a mile or more until there is rain and, with it, a shattering of the calm

> *Sound of the forest and mountain:-*
> *When forests do murmur and mountains do roar*
> *Close all your windows and bolt up your door*

b) Echo on the shore

The shores sounding in a calm, and the sea beating in a murmur or an echo clearer than usual , are signs of wind and rain to come

c) Ringing of bells

The ringing of bells is heard at a greater distance, with the sound coming and going, before rain

10) ... See stars?...

Sun, moon, haloes and what's in the stars!

The sun first, supreme creator of weather!

Consider the following...

a) The sun's tidywork – see the clouds that it creates!

This is the story:-
    The sun's heating of the ground forces air to rise , creating winds that move vertically. thence horizontally. Clouds and cloudbanks form as a result. Once, however, the upper winds can venture no further but are then forced to descend, the air will begin to clear of cloud in that region But that isn't the end of the story. There are knock-on effects!

b) Clouds 'brushed away' by the sun?

    While heat will act to evaporate cloud, there is the more obvious role of the sun in first making cloud from the ever- rising currents of air where ground is heated and that then set off on

51

their own adventure; thus a merry-go-round!

c) When hot ground changes the weather,
        creating majestic storms

The hot ground of *(in particular)* a typical British heatwave is always a challenge to the stability of any fine summer weather. Increasingly large thunderheads will show that the sun's work is drawing air sufficiently high for cooling and condensation to take place, very often on a majestic scale *with majestic storms to match!*

d) Weak sun – what then?

Weakly -angled winter sunshine will have no direct effect on the weather overhead and on what may or may or may not fall to the ground. Weather systems that are themselves products originally of solar heat will do this

e) Solar spotlight

The sun as a spotlight on cloud formation is able to trace cloud patterns that the weather-wise will identify

f) Mock suns

Reflections may abound as the sun picks up an advancing cloud layer heralding rain to come

g) Pale, watery, on rising

      This is the result of it spotlighting and being dimmed by upper level cloud associated with an advancing rainbelt. The weather will deteriorate by degrees, with rain commencing within a few hours

h) Shining as though through ground glass

      Identical message – rain coming!

i) Pale yellow, or red, sun on rising

      a wet day ahead, rain starting soon

j) Bright yellow, on rising

      wind strength increasing, possibly rain also

k) Sun with halo (detached ring)

      rain coming,, likely after several hours

Lunar influence, the moon has her say

A feast of weather lore surrounds the *moon*. From the proven, meteorological, point of view - and leaving aside the moon's well known pull on the tides - her role is to reveal cloud formations capable of producing rainfall, as does the sun by day.

There is no evidence that the weather obediently adjusts to every change of the regular lunar cycle... but...

If the moon rises haloed round
Soon we'll tread on deluged ground

Indeed, yes! The saying refers to the moon with a detached ring, therefore, the halo, not to the (generally broader **attached** circle caused by a low mist and which is typical of a quiet weather period of the autumn... the so called 'harvest moon' ... 'that never filled a pond'.

a) Detached ring (halo)

As with the sun, a detached ring highlights an advancing sheet of *altostratus* cloud *(following from cirrostratus)* that will thicken, become progressively lower and lead to a spell of continuous rain for several hours
.

b) Sharp, with horns

When the moon's horns are sharp and well defined expect rain the following day or, in winter, frost before rain

c) Reddish moon, through haze

When the moon glows reddish brown through the haze the weather will stay clear

If the full moon rises pale, expect rain.
    … advice that will apply to her other phases

> *Last night the sun went pale to bed*
> *The moon in haloes hid her head,*
> *The boding shepherd heaves a sigh*
> *For, see! A rainbow spans the sky....*
>
> *from a poem by Erasmus Darwin*

d) When seen by day

If the moon is seen by day the weather will be cool

On hazy days when the weather is settled, the moon would be concealed.

e) Once in a blue moon

For sure, , this is very rare, perhaps a once-in-a-lifetime experience. The cause is dust, usually from a desert region, taken to a high altitude. More likely than not, the weather will be fair at the time

f) For the lunar test-bench

Two full moons in a calendar month bring on a flood
Bedfordshire saying

To see the old moon in the arms of the new could be a sign of bad weather to come... but in Suffolk is reckoned to indicate fine weather, as is the turning up of the horns of the new moon
(based on the supposed clearance of clouds said to
take place when the full moon rises)

The weather is generally clearer at the full than at the
other stages of the moon, but in winter the frost is sometimes
more intense..   *Lord Bacon*

The full moon eats clouds
Nautical saying

Near full moon, misty sunrise,
Bodes fair weather and cloudless skies

If the full moon rise red, expect wind

A new moon's mist
Will never die of thirst

Threatening clouds without rain in the old moon indicate drought

If the moon show a silver shield
Be not afraid to reap your field -
(Reference to the misty 'harvest moon

g) Lucky stars

When the sky seems very full of stars expect rain the following day or, in winter, frost

From the joy of a brilliant night- sky prepare for the ground to cool rapidly after the demise of a typically north-west wind that leads to frost in due season. Closed-in valleys will prove the coldest spots locally

With regard to the clarity of the sky, this will show that the Earth's atmosphere at the point observed is unlikely to be static and that change of weather is 'the name of the game'. Dry it may be, if just for the night

Did it really happen, was I really there?
Was I really there with you? -
Yes, we lived our little drama
And stars fell on Alabama that night

from the song by Mitchell Parish and Frank Perkins, 1934

…a spectacular shower of meteors was observed at Alabama in August, 'the night the stars fell'

]

11)   .. Waterspouts?....

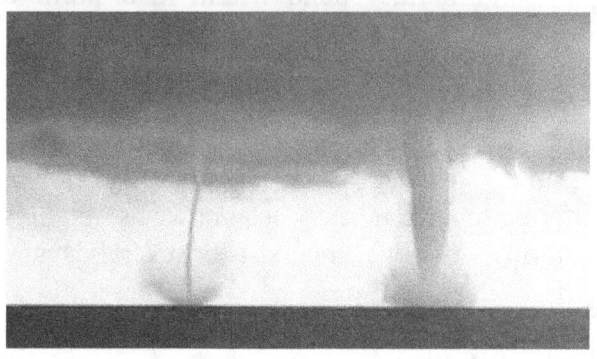

a) Agitated sea, no wind

In the absence of wind a long, rolling swell indicates
that a storm is close

b) East or west coast of Britain

An east coast ripple known as a 'twine' is said to precede
a gale from the south-east.   On the west coast a heavy surf is held
to be the forerunner of a storm
c) Sea foam

When the foam of the sea retreats it is said in Scotland to
be 'leukin for mair' and that stormy weather is at hand

d) Surface of the sea in harbour

When the surface of the sea in harbour remains calm but yet there is a murmuring sound within it, a storm will break within a few hours

e) 'Sea lungs'

Glittering foam in a heavy sea known as *'sea lungs'* is a sign that a stormy weather will persist for several days

f) White circles of froth

White circles of froth or of bubbles of water on a calm and smooth sea denote wind or, if continuing, a major storm

g) Waterspouts

Waterspouts are not produced in cold weather but when the atmosphere is close and thundery . Aristotle

h) Heaving and sighing

Before a major storm the sea will 'heave and sigh'
Admiral Fitzroy

i) River foam

Much foam in a river foretells a storm *Scotland*

j) River flood

> *A river flood*
> *Fishes' good*
>> *Spanish proverb*

k) Rising, no rain

If the river Tweed rises without rain, the same will not be long delayed.   Local saying

12) ...Or a White Christmas?

a) Unfinished business

For those who study the seasons and evaluate their progress, business is never finished and satisfactory explanations often at a premium...

b) Room for disagreement…

.a life dependent on the seasons is what the almanacs all agree upon, if not on what will happen...

A green Christmas,
A white Easter?

Flowers in May,
Look out for your hay?
(it will be a poor summer)

or a frosty May? ...

Always a setback at the time, ground frost in May is considered better than too early a start to the summer and for which there is a price to be paid in the opinion of most. But not every time

c) Hard winter, hot summer?

The, late 1940's saw this belief vindicated; the early 1960's saw it turned on its head. In '63 there was no exceptional summer to crown the earlier discomforts of a freezing winter

Snow on Christmas Day?   A safe bet?   Not so for Britain as a whole where the Christmas holiday interval so often sees a break in the unsettled pattern and turns out mild and sunny

A warning.   There is evidence that if deep winter cold comes early at around mid-December, then the winter will continue that way, frosts harden, snow drowning roads and hedges... *Long notice, long last..*

d) Await the thaw!

A really firm cold spell isn't easily shifted, nor should one expect an early end to drought in an occasional Continental-type summer with its record high temperatures. Nor, for that matter, the hasty retreat of a rainy influence.   Nature takes her own time

e) Swithin's 40 days of either sun or rain?

Never mind the legend that arose through St Swithin making his presence felt in the year 963 at Winchester for not being buried where the 'sweet rains of heaven' might fall upon his grave. To a limited extent there is a 'persistence of type' in British weather. Thus once a particular spell has set in, there is little immediate chance of any change.

Swithin had got it about right ... but in the end it may depend on where one takes one's observations. For the record, on July 15th:-

> *St Swithin's Day, if ye do rain*
> *For forty days it will remain;*
> *St Swithin's Day, and ye be fair*
> *For forty days 'twill rain nae mair.*
> *(Quote from the Scottish version – they have it there as well)*

When the saints come marching in...

While not equally remembered, many of the saints have weather connections tied to their respective festival dates:-

a) St Vincent, Jan 22

If St Vincent has sunshine
One hopes much rye and wine

(against that, too early a sun in January was to be deplored)

b) St Paul, Jan 25

St Paul's Day be faire and cleare,
It doth betide a happy yeare

(The last twelve days of January are said to rule he weather for the year)

c) Candlemas, Feb 2

At Candlemas Day
Another winter is on his way ... alternatively

If Candlemas Day bring clouds and rain,
Winter is gone and won't come again

d) St Valentine, Feb l4

To St Valentine spring is a neighbour

e) St Peter, Feb 22

The night of St Peter shows us what weather
we shall have for the next forty days

f) St Matthias, Feb 24

> St Matthias breaks the ice, but if he does not
> find any he will make it

*(A dry and cold March never begs its bread) – A load of March
dust is worth a king's ransom)*

g) St David, March 1

> *Upon St David's Day*
> *Put oats and barley in the clay*

h) St Patrick, March 17

> *St Patrick's Day, feel the warm side of a stone*

i) St Joseph, March 19

> *St Joseph's Day clear*
> *Allows a fertile year*

j) St Benedict, March 21

> *Sow thy peas or keep them in thy rick*

k) Lady Day, March 25

> *If't on St Mary's bright and clear*
> *Fertile is said to be the year*

(The 'Blackthorn Winter'... when warm days end-March or in early April that bring the blackthorn into bloom are likely to be followed by a renowned cold period)

When April blows his horn
'Tis good for hay and corn

l) St George and St Mark, April 23, 25

At St George the meadow turns to hay

When St George cries 'Go'!, St Mark cries 'Hoe'!...
but ... trust not a day
Before birth of May

m) Sts. Mamertius, Pancras, and Gervais May 11,12,13

Who shears his sheep before St Gervais loves his wool
more than his sheep

n) St John, June 24

If it rains on St John's day it will rain seven weeks

o) St Swithin, July 15 (previously quoted, page   )

When it rains at St Swithin he is christening the apples

p) St Vincent, July 19

At St Vincent the rain ceases and the wind comes

q) St Jacob, July 20

Clear on St Jacob's Day, plenty of fruit

Dry August and warm
Doth harvest no harm

r) St Margaret, August 13

The proverbial flood of the day is considered to be
good for the harvest in England

s) St Bartholomew, August 24

If this day be misty, cold weather will soon come;
At St Bartholomew comes the cold dew

t) St Matthew, September 21

St Matthew's Day bright and clear
Brings in good wine next year

u) St Luke, October 18

St Luke's summer
The belief in a mid-October burst of late summer is well
founded, if of late beginning somewhat earlier than this
date

v) St Martin, November 11

Expect St Martin's summer, halcyon days

w) St Thomas, December 21

As the weather cock shows on this day, so will the wind be for the next lunar quarter

The mostly fun part

CONTENTS   PART TWO
What they say about the weather

Page

What's the weather? - ask:-.

1) ...the birds and the bees...

a bird doesn't even need to stop and think,
- unless he's an owl!

Swallows
Swallows high
Staying dry;
Swallows low
Wet 'twill blow

Larks   If larks fly high and sing long, expect fine weather

Rooks   If rooks feed in the streets of a village, a storm is approaching

When rooks drop in their flight as if pierced by a shot, it indicates rain

Rooks will not leave their nests on a morning before a storm

Ravens   Ravens, when they croak continuously, denote wind, but rain if the croaking is interrupted or stifled

When ravens sit in the sun, it will stay fine

If the raven in winter makes several different cries, it is the sign of a storm

Crows   The continual prating of the crow, chiefly twice or thrice, quick calling, indicates rain and stormy weather

Blackbirds   When the voices of blackbirds are unusually shrill, rain will follow

Robins   If a robin sings on a high branch of a tree it will be fine weather, but if one sings near he ground, the weather will be wet

Thrushes   When the thrush sings at sunset, a fair day will follow

Woodpeckers When woodpeckers are much heard, rain is
certain

Sparrows If sparrows chirp a great deal, it will be wet

Fowls If fowls huddle together outside the henhouse
instead of going to roost; expect rain and wind;
Likewise, if they grub in the dust or clap their
wings

Cocks If the cock goes a-crowing to bed,
He'll certainly rise with a watery head

If cocks crow during a downpour it will be
fine before night

Guinea-fowls Guinea-fowls squeal more than usual before rain

Land birds Land birds are observed to bathe before rain

Tree birds -

Heron If birds that dwell in trees return eagerly to their
nests and leave their feeding ground early, it is a
sign of storms, but when a heron stands
inattentive and melancholy, it denotes only rain

Herons in the evening flying up and down as if
doubtful of where to rest point to rough weather

Geese   *The goose and the gander*
                     Begin to meander;
                     The meaning is plain
                     They are dancing for rain

Ducks and Geese   When ducks and geese fly backwards and
                      forwards, continually plunge into water and
                      wash incessantly, it is a sign of early rain

                      If the wild geese gang out to sea
                      Good weather there will surely be

Gulls   Seagull, seagull, sit on the sand
                     You bring bad weather flying o'er land

Island birds When, in summer, many birds that dwell on
                  an island turn up in flocks on the mainland,
                  it indicates rain

Peacocks   When the peacock loudly calls
                     Soon there'll be both rain and squalls

Owls   Screech owls are most noisy just before rain

                     An owl hooting quietly in the rain indicates an
                     improvement shortly

Petrels   When petrels gather in their numbers at the
                     stern of a ship, storms are near. The petrel has
                     long been a token of an impending tempest

Turkeys   Turkeys perched on a tree and refusing descent
                     indicate snow

Birds silent

When birds are unusually silent in the summer,
electrical storms and thunder are close

Bees ...

a) 'Staying put'   When many bees enter a hive and none leave,
rain is close

b) Not swarming   Bees will not swarm
Before a near storm

c) Knowing   A bee was never caught in a storm

If bees stay at home
Rain will soon come

d) Wide- ranging   *Days are warm and skies are bright*
                     When bees to distance wing their flight,
                     But if patrolling close to home
                     Prepare for rain or storms to come

2) ...or obliging animals...

a) **Crowding** when animals bunch together, expect rain

b) **Dogs** The unusual howling of dogs portends a storm

Dogs making holes in the ground, howling when someone goes out, eating grass in the morning or refusing meat point to rain or snow

If dogs persist in rolling on the ground, or become drowsy or stupid, it will rain

c) **Spaniels etc.** It will be wet if spaniels sleep more than usual or when a dog curves its belly to the ground

d) **Cats**   When a cat sneezes it is a sign of rain

Before rain or wind a cat will scratch the wall, or a post or table leg.

A cat will sit with its back to the fire before snow

When a cat licks herself and turns her face, the wind will blow from that direction

A cat apparently electrified, fur raised, tail up, will indicate (if not an approaching dog) an imminent rise of wind, and for certain a weather change if unusually playful or quarrelsome

e) **Cat washing**   The cardinal point to which a cat turns her face after rain, when washing, is the direction from which the wind will blow

When cats wipe their jaws with their feet or put their paws over their ears, it will rain

d) **Horses**   If horses stretch their necks and sniff the air, rain will ensue

Horses sweating in the stable is a sign of rain

If horses are more than usually restless and uneasy, or assemble in the corner of the field with their heads to leeward, it will be foul weather

If young horses rub their backs against the ground, the weather will change; and, again, if lively without apparent cause.

e) **Asses**   *Hark. I hear the asses bray*
                 We shall have some rain today!

donkeys   Donkeys have been known to prick their ears
                 forward or rub against walls before rain

                 Be sure to stack your hay and corn
                 If the donkey blows his horn

f) **Cattle**   When cattle lie down during light rain, it will
                 soon pass, and once remaining on hilltops fine
                 weather has come

                 When cows fail their milk expect unsettled and
                 stormy weather. In winter, if they bellow in the
                 evening, expect snow, or rain if lying with their
                 heads upon the ground or, in summer, slapping
                 their sides with their tails

g) **Bulls**   Bulls lick their hooves or kick about before rain

h) **Oxen**   Oxen have been noted to lie on their left side
                 before a change to rainy weather

i) **Goats**   Goats leave high ground and seek shelter before
                 a storm, but if goats and sheep quit their pasture
                 with reluctance it will not rain before next day

j) **Sheep**   Sheep may turn their backs to the wind before
                 the onset of rain, or they may gambol and fight
                 and attempt to box each other. Older sheep may
                 eat greedily before a storm

                 If sheep feed uphill in the morning it is a sign
                 of fine weather

k) **Pigs**   Pigs running unquietly, or up and down with hay or litter in their mouths foretell a storm

When pigs carry sticks
The clouds will play tricks;
When they lie in the mud
No fear of a flood

l) **Beavers**   In early and long winters the beaver prepares his house one month earlier than usual

m) **Mice**   If mice seem to frolic or run about more than usual, rain must be expected

n) **Moles**   Moles plying their works and undermining the earth point to regular rain but, in summer-time, respond to heat by forsaking their trenches

o) **Squirrels**   When squirrels lay in a large supply of nuts, expect a cold winter, but -
When they eat them on the tree
Weather mild as mild can be

p) **Weasels, stoats**   If running about briskly during the forenoon there will be rain later that day

q) **Bats**   If excitable, crying or flying into the house there will be rain within hours; otherwise the presence of many bats accords with settled weather

r) **Hedgehogs** These creatures conceal themselves before a
major change in the weather

3) ...or insects...

The early appearance of insects points
to a good spring and good crops

a)   **Ants**   Ants withdraw into their nests and busy
themselves with their eggs before a storm, and
also while building their walls

Expect stormy weather when ants travel in lines

An open ant-hole speaks of clear weather

If in July ants are busy building and enlarging
their piles, an early winter will follow

b)   **Wasps** Wasps building their nests in exposed places
indicate a dry season. If numerous and busy
they point to fine weather

c)   **Hornets**   build their nests high in warm summers

d) **Spiders**   Spiders work harder and spin their webs in
advance of an increase of wind, as they cannot
do so later

If spiders are totally indolent, rain soon follows

If spiders wait till eve before changing their webs,
expect a fair, dry night

Spiders crawling on walls more than usually
prognosticate rain, also when they break off and
remove their webs

If a spider works during rain the weather will
soon clear
When the spider cleans his web, fair weather is
indicated

If the filament of a spider's web is long, expect
a fine day

When spiders' webs in air do fly
The day will soon be very dry

Spiders in motion indicate rain while webs floating in an autumn sunset point to frost

e) **Woodlice**   Woodlice in great numbers, long wet spell

f) **Flies**
A fly on your nose,
You slap, and it goes;
If it comes back again
It will bring a good rain

(this is a reliable saying as increasing damp before rain causes insects to be more than usually troublesome)

g) **Butterflies** The early appearance of butterflies points to settled weather

h) **Fireflies**   Fireflies in great numbers indicate fair weather continuing

i) **Gnats**   Gnats playing up and down tell of heat

j) **Crickets**   Continued chirping of crickets warns of rain

k) **Clock beetle**   Flying circularly and buzzing, the clock beetle gives an assurance of fine weather

4) ...or reptiles, amphibians...

a) **Snake trails**   Trails may be seen more readily before rain, which is the time snakes are most active

Reptiles passing the winter semi-comatose show signs of weather changes by a response in their attitude

b) **Frogs**   When frogs warble, they herald rain...the louder the frog, the more the rain

When frogs spawn in the middle of the water it is a sign of drought

Tree frogs piping during rain indicate a continuance

The green tree-frog is noisy before rain

c) **Toads**   See toads in great numbers, and rain will soon fall

5) ...or would you rather trust a fish? ...

(whales , etc. included in this section)

Generally, in rivers, lakes   *Fish bite the least*
                               With wind in the east

a) **Pike**   When pike lie quietly on the bed of a stream
                   rain will follow

b) **Clam beds** Bubbles over clam beds indicate rain

c) **Eels**   It is a sign of rain if eels become very lively

d) **Before rain**   Fish bite readily and swim near the surface
                   before rain

e) **Mullet**   Mullet run south at the approach of a cold
                   northerly wind

f) **Bass**   Bass leave shoal water up to 24 hours before
a thunderstorm

g) **Black fish** Black fish in schools indicate an approaching
gale

h) **Porpoises, whales**   When spouting about ships at sea,
storms are imminent

Porpoises in harbour indicate a coming
storm, and when swimming to windward
or running into bays and around islands

i) **Dolphins** Dolphins sporting in a calm sea indicate wind

6) ...or a plant or a tree?...

a) **Dandelion, pimpernel** The dandelion and scarlet pimpernel are among a group of wild plants that close their petals before rain or evening damp

b) **Chickweed** Chickweed expands its leaves boldly and fully for fine weather

c) **Clover** Becomes rough to the touch if storms approach

d) **Anemone** The yellow wood anemone droops before rain

e) **Trefoil** The stalk of the trefoil swells before rain

f) **Colt's foot** If the down of colt's foot, thistle and dandelion begins to fly when there is no wind, rain will follow

g) **Hips, haws** *Many hips and haws*
Many frosts and snaws (Scotland)

h) **Nettles** Many dead nettles late in the year is the sign of a mild winter

i) **Hawthorn** It is always cold when the hawthorn blossoms

j) **Marigold** If the African or Cape marigold open early in the day, the weather will stay fine

Cape Marigold

Trees   *When leaves show their undersides*
                  Be very sure that rain betides

a) **Silver maple**   The silver maple shows the lining of its
                  leaves before rain

b) **Aspen**   The leaves of aspen will tremble before the
                  approach of thunder

c) **Lime, sycamore, plane & poplar**
                  Before rain their leaves show more of their under
                  surfaces

d) **Dead branches**   A cracking sound of dead branches, or
                  if they snap and fall to the ground when
                  it is calm, rain is due

e) **Ash & oak**   *When the oak's before the ash*
                  Then you'll only get a splash;
                  If the ash precedes the oak
                  Then expect a summer soak*

(* there are a number of sayings relating to the budding of
these two trees, some contradictory, and in any case the
ash is habitually late to reveal its green)

f) **Leaves & straws**   When these play in the air yet there is
                  no breeze, a wind will soon arise. A
                  small swirling of sand around a heap
                  may be noted in similar conditions

7) ...or a suffering human?...

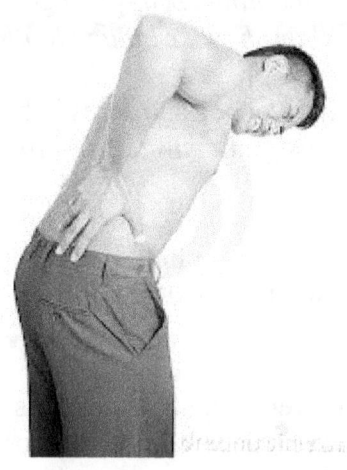

a) **Pains in the joints**

When rheumatic people complain of more than
ordinary pain in the joints, it will rain

b) **Motes**   The deceptive appearance of motes before
the eyes is said to presage rain and storms

c) **Uneasy**   Persons of a weak or irritable constitution are
uneasy before storms

d) Shooting corns

*'A storm your shooting corns presage
And aches will throb, your hollow tooth will rage'*

If corns, wounds and sores itch or ache more than usual, rain will fall shortly

e) Ears ringing A ringing in the ears may indicate a change in the weather and an increase of wind

f) Appetite When everything at the table is eaten it will more than likely point to dry, breezy, mostly settled weather

g) Nocturnal Some will experience discomfort at night if wind and weather are about to change

'Old sinners have all points of the compass in their bones and joints!'

8) ...or look round and about? ...

a) **Chimney smoke**   If smoke from chimneys first descends across the eaves of a building before rising, expect rain

b) **Clear sound**   A still dull day when distant sounds are clearly heard speaks of rain to come

c) **Watercress beds**   When watercress beds steam on a summer night the next day will dawn fine and sunny

d) **Seaweed**   A piece of kelp or seaweed hung up and showing damp by day will point to rain

e) **Cream and milk**   The unexpectedly early sourness of cream or milk indicates thunder

f) **Whistling**   Whistling when at sea, is credited with the raising of a contrary wind

g) **Damp walls**   Cottage walls exuding damp indicate rain

h) **Catgut**   Catgut and whipcord untwist and become longer during a dry state of the air - thus the principle on which a 'weather house' is created

i) **Ropes**   When difficult to untwist, rainy weather is likely or is present already. *(A sailor will note the tightening of his cordage as a sign of bad weather)*

j) **Camphor gum**   When dissolved in alcohol, the feathery crystals created will rise before rain

9) ...or tap a barometer?...

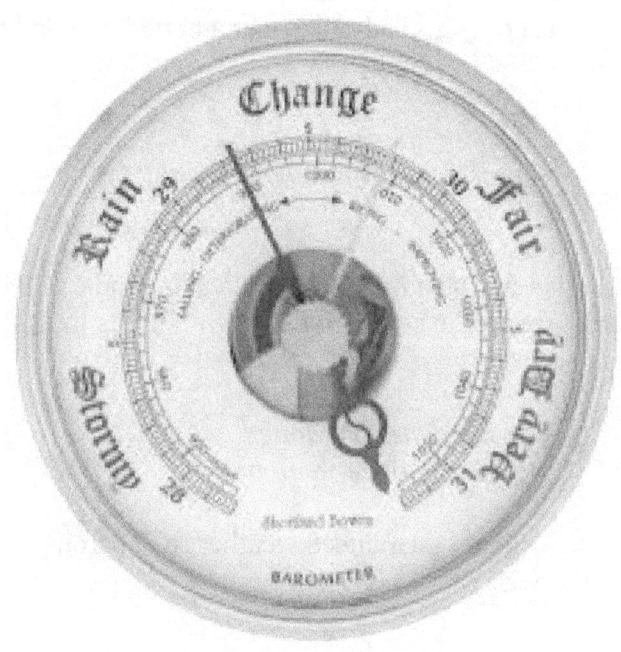

Barometer indications ...

Low or high

Falling low,'
Prepare for a blow;
Rising high,
Mostly dry

Warnings:

(1) Too sudden a rise after falling is likely to lead to a further fall. Trust only a steady rise over a period of a day or two. Without that assurance:-

> First *rise after low*
> *Foretells a stronger blow*

(2) A higher and steady barometer reading indicates dry, warm weather in summer, but in winter it is likely to be cold

> *Long foretold, long last,*
> *Short notice, soon past..*

.... the virtually guaranteed weather rule, barometer-wise
or otherwise

a) Low, after rain

Any clearance of the sky will be short-lived

b) Falling, winter and summer

The barometer falling with a southerly or westerly wind indicates damp weather, stormy, mostly mild in winter; stormy, mostly cool in summer

c) Falling, wind backing*

Always a sign of rain, snow or storm, according to extent and season (* anticlockwise change of direction)

When the wind backs and the weatherglass falls,
Your signal for certain of rainfall and squalls

d) Falling for wind and rain

The barometer falls lower for high winds than for heavy rain. The steeper the fall the greater the gale

e) Rising, wind veering*

Await a change to drier conditions
(* clockwise change)

f) Rising, wind southerly

Weather improving, warmer

g) Rising, wind easterly

Very hot in summer,
Cold in winter and dry for a time, but a change to south-east to south will bring snow before any thaw

h) Sudden, jerky movements

Squally weather, variable temperatures

10) ...or get canny with the seasons?

1) Early Autumn Can be nice!	
2) Late Autumn – Early Winter Gales	
3) Late Winter – Early Spring Possible extremes	
4) Late Spring – Early Summer Can be a surprise!	
5) Late Summer – Early Autumn Ups and downs	

Thus five seasons, not four ...

1) **early autumn,**

    mid-September to mid-November...
    mostly mild, with a handful of deceptively brilliant
    days that speak of summer not yet past. But otherwise
    showing a liking for cloud and rain, especially around the
    late-month periods.
    (Mid-October, *St Luke's Summer.* Mid-November. *St
    Martin's Summer)*

2) **late autumn to early winter,**

    late-November to mid- January...
    stormy, with just a few sunny breaks in which to recover
    one's senses and including an almost regular one around
    Christmas. The period as a whole more often mild than
    cold, the chief hazard being flooding

3) **late winter to early spring,**

    mid-January to mid-April...
    featuring one of two main weather types, either mostly
    mild, wet, galey at the start, but gradually easing;
    alternatively, and rarer - diving into cold with east wind,
    frost and snow. Night frosts at any time

4) **late spring to early summer,**

    late-April to mid-June...
    featuring normally a sharp decrease in rainfall at the start
    accompanied by a wide temperature range that may
    include night frosts until mid-May. A foretaste of full
    summer is a fairly regular feature towards the end

5) late summer to early autumn,

late-June to early September...
featuring an increase in rainfall, but in one or two years in
ten an exceptional dry season with its accompanying
deficiency of water. This said, thundery rainfall intervals
may produce flash flooding locally while drought persists
generally

NOTES

NOTES

NOTES

NOTES

Also available through Amazon

www.ingramcontent.com/pod-product-compliance
Lightning Source LLC
Chambersburg PA
CBHW060351190526
45169CB00002B/567